Ologies

Four Essays

Ologies

Four Essays

Chelsea Biondolillo

Etchings Press
Indianapolis, Indiana

This publication is made possible by funding provided by
the Shaheen College of Arts and Sciences and the English Department
at the University of Indianapolis. Special thanks to the student who judged,
designed, and edited this chapbook: Tierney Bailey, Mikaela Bielawski, Rachel
Holtzclaw, Alyssa Kauffman, Adam Kuhn, Zach Lee, Dani McCormick, Rachel
Neawedde, Mirna Palacio Ornelas, and Jessi Tillman.

UNIVERSITY *of*
INDIANAPOLIS.

Published by Etchings Press
1400 E. Hanna Ave.
Indianapolis, Indiana 46227
All rights reserved

etchings.uindy.edu
www.uindy.edu/cas/english
Printed by Ingram Spark

Published in the United States of America

ISBN 978-0-9903475-2-1

23 22 21 20 19 18 17 16 15   2 3 4
Second Printing, 2019

Table of Contents

PHRENOLOGY // an attempt

In the beginning, science class was exhilarating. Not yet acclimated to the squeamish hallmark of my sex, I would pick up anything.

That delirious green beetle I found desiccated on the windowsill had a back so soft that my fingertip wasn't sensitive enough to appreciate it. I held it to my eyelid, finally my lips. I rubbed one spot, over and over, until the softness had left it.

I touched an electric fence once[1] when I was ten, at a friend's house. The voltage was low enough that it thrummed through me, like a sine wave from bones to tendons to veins. I was not deterred at all, and thought often, later, of getting that feeling back: the hum of electrons coming together.

My dearest Esther,—& all my dears to whom she communicates this doleful ditty, will rejoice to hear that this resolution once taken, was firmly adhered to, in defiance of a terror that surpasses all description, & the most torturing pain

Later, at the beach, tiny jellyfish washed up in daunting numbers under windy skies, looking for all the world like egg whites, emptied of yolks. I wanted to hold one, but my grandmother said they had a sting like an electric shock. As soon as my grandmother turned her back, I reached out a tentative finger. The mass had a solid feel; it had an unexpected skin around the

---

1   *A lot of words in English confuse the idea of life and electricity, like the word livewire.* Laurie Anderson

clear softness. There was no shock at all.

Differentiation. Orogeny. Anion. Mitosis. Batholith. Telophase. Syngamy. Escape velocity.

Things separate from the whole: atoms, stars, continents, stones, leaves, cells, skin, parents. Things also come together.

It isn't a good idea to be the girl who likes torn skin and spiders and snakes. You might find yourself differentiating from the group. Stratifying.[2] If you are the girl who looks when a bug is stepped on or a mouse is half eaten by your own cat, your orbit may widen out and flatten, like a comet, your view moving farther from the center to an outer perimeter. It will feel as far away as Cassiopeia, as Arcturus.[3]

When the dreadful steel was plunged into the breast - cutting through veins - arteries - flesh - nerves - I needed no injunctions not to restrain my cries. I began a scream that lasted unintermittingly during the whole time of the incision - & I almost marvel that it rings not in my Ears still!

Sometimes an obsession feels like looking out different windows and seeing the same landscape, like turning the radio dial and hearing the same song over and over.

Let's face it, no one ever believed I was going to be a scien-

---

2   *We shall of course be told that they must go into the water before they can learn to swim; but what is proposed is not to teach them to swim: it is to throw them all at once into a fathomless ocean, where they will drown themselves, and pull down those who were swimming there, or trying to swim before them.* Francis Parkman, "The Woman Question" 1872

3   *If you can't give me poetry, can't you give me poetical science?* Ada Byron, Countess of Lovelace

tist: I lacked focus and drive, but still my mother bought me the Gabriel Tri-Lab-Pak for Christmas, 1983. She would not allow any of the red-capped chemical bottles to be opened after reading the warnings on them. I cracked open the sulfur bottle, from time to time, having never smelled a rotten egg, but I left the rest closed. Instead, I dutifully performed identification streak tests on the non-volatile ceramic tile, and scratched pennies with the mineral specimens—to rate their hardness against the copper. The glossy silver hematite scratched a surprising rusty line; the quartz gouged the penny like a surprising root into the thin membrane of a shin. Suddenly the world seemed full of shifts, full of things that were more than they appeared to be, harder than they seemed.

He said you mustn't drop the tarantula, that if you did, its belly would burst open like a tiny watermelon. We all pictured the tragedy of that ghastly undoing, then waited our turns. The man with the spider placed it carefully on my stiff, upturned palm. He was all nerves, explaining to the rest of the class the importance of moving slowly, of being serious, of being calm.

It felt ungentle, unlike a hamster or woolly bear caterpillar in every way. I concentrated

When the wound was made, & the instrument was withdrawn, the pain seemed undiminished, for the air that suddenly rushed into those delicate parts felt like a mass of minute but sharp & forked poniards that were tearing the edges of the wound – but when again I felt the instrument – describing a curve – cutting against the grain, if I may so say,

on the rhythm of one leg—bristled like a pipe-cleaner then tapering to a point—lifting then stepping, as the weight of its fragile abdomen shifted with each step. I was too focused on not dropping it to be afraid.

Later that year, someone came into the first grade classroom and squeezed all of the monarch butterfly co-coons that were hanging in a row above the art table. It had to be an older kid, because we couldn't reach them. What came out of their onion-skinned green shells looked neither like what went in or what was supposed to emerge. They were still blended, a mix of land and air, when they were split open like grapes.

Most kids wouldn't hold the fire-belly salamanders after the herpetologist explained that their brightly colored skin contained poisonous alkaloids, which tasted bad to predators. All I saw was that hysterical orange stomach, and I greedily stretched out my cupped hands to take the squirming wetness.

while the flesh resisted in a manner so forcible as to oppose & tire the hand of the operator, who was forced to change from the right to the left – then, indeed, I thought I must have expired.

I rested my thumb on its back, to still it in my hand; I watched its toothless mouth open and its lashless eyes blink. Later, in the darkened classroom, as the class watched a film on wetlands, I scratched deep grooves in my thighs. The hives were palm-shaped with raw finger-stripes, where my hands had rested as the film began. The best part, I remember, was how my hands didn't itch, only the skin they touched—like I was the poisonous one, now.

I am looking for a way to get at the experience of a thing,

the memory of it, to better understand the meaning of it. I am expecting the process to be messy.

In the road once, in the fourth grade, after getting off the #10 city bus to Harold Street, I passed close by an opossum, torn open by tire treads. I knew it was an opossum, and not a cat, though much of its distinguishing characteristics had been ground into the pavement, because of its pink tail. Though I walked fast, and I knew it was wrong to look,[4] I still saw a movement among all the red and pale slicked shapes, between two ribs poking up. They were pink like her tail, and also hairless, and they squirmed blindly: thirsty and shocked at the sudden cold.

I attempted no more to open my Eyes, - they felt as if hermetically shut, & so firmly closed, that the Eyelids seemed indented into the Cheeks.

I didn't want to do the catching or crushing or cutting, ever. Destruction was never the point. I wanted to see the aftermath; I wanted to know how things were unmade.

I spend an afternoon trying to find a woman pioneer in the field of electricity, but end up poring through reports of

---

4   *In an address to the American Society for the Prevention of Cruelty to Animals on Thursday afternoon, President John P. Haines made the following statement, referring to the killing of cats by teachers in the public schools for purposes of dissection: "Women who practice it get fond of it, and women who get fond of cat killing are on the high road to homicidal mania... It has been asked why this tendency should be so stressed in reference to women. I reply that women are, to a large degree, the creatures of impulse."* New York Herald, Feb. 17, 1900

women who have been executed in the electric chair.[5] *It often happens on accident like this, I don't go looking for the split open things or the dark insides.*

The first one was in 1899. By then, they'd figured out the voltage, and she went fairly quickly.

What would a phylogeny of that which terrifies look like? How can I create a taxonomy for all my dark fascinations?

In 1981, I was about to turn seven. My grandparents had taken

The instrument this second time withdrawn, I concluded the operation over - Oh no! presently the terrible cutting was renewed - & worse than ever, to separate the bottom, the foundation of this dreadful gland from the parts to which it adhered -

me on a birthday trip to the beach almost two weeks early that year—no one remembers why. Maybe my mother and new stepfather had a trip planned for Memorial Day weekend; maybe my new baby half-sister was sick. But on May 18th, a Sunday, I was in a small five and dime store on the Oregon Coast, worrying over a teddy bear. I was allowed to pick one as a birthday present, from what seemed like hundreds, each different from the others.

There was a very soft smallish bear that I liked a great deal,

---

5 *My dear Mr. Governor: Please forgive me for bothering you...I have been paralyzed for more than three years and I could not look after Gennie as I wants to. I know she done an awful wicked thing when she killed Miss Belote and I hear that people at the penitentiary wants to kill her. But I am praying night and day on my knees to God that he will soften your heart. If you only save my child who is so little, God will bless you forever.* Charlotte Christian in a letter to William Hodges Mann. Virginia Christian, age 17, was electrocuted on August 16, 1912

but it seemed silly to waste this amazing opportunity (ANY! BEAR! PICK! ANY!) on something so small. The biggest bear was not soft, but had matted dark brown fur, like a toy from a county fair. It had a yellow and black nose. It was almost as big as me. I picked this bear, the largest one, even though I remember very clearly liking the small one more. The remorse was already upon me when we walked out into a rare sunny day and my grandmother pointed up at the sky, at the growing white cloud of a volcanic eruption.[6]

For days, back in town, I watched the cloud get taller. The Portland weather conspired, staying unusually clear of low, gray mist. The ash cloud looked white, though I knew from the television that it was towering over dark mud and seething pyroclastic flow. Dead trees, dead deer, dead scientists, walls and roofs and stones, were all rushing over the banks of the Toutle River under that ghostly white shape. I nearly went blind from staring at it, trying to catch it moving. But it always looked perfectly still.

While watching the Brakhage film, *The Act of Seeing with*

---

6   *The eruption cloud is very solid-looking, like sculptured marble, a beautiful blue...darkening towards the top—a wonderful color. One is aware of motion, but (being shaky, and looking through shaky binoculars) I don't actually see the carven-blue-sworl-shapes move...It is enormous. Forty-five miles away. It is so much bigger than the mountain itself. It is silent, from this distance...It looks not like anything earthy, from the earth, but it does not look like anything atmospheric, a natural cloud, either... the shapes are far more delicate, complex, and immense than stormcloud shapes, and it has this solid look; a weightiness, like the capital of some unimaginable column—which in a way indeed it is, the pillar of fire being underground...To us it is cataclysm and destruction and deformity.* Ursula LeGuin describing the 1981 eruption of Mt. St. Helens in her journal

One's Own Eyes, I made the following notes:

Technicians silently split bodies open. One attends the cadaver of a young woman. Her breasts are still soft and full; they move, as they would if she were dancing. He cuts and lifts her skin from her ribcage; he pulls with firm movements, deliberate. The scalpel saws gently and she is opened like a suitcase.

And I think of Fanny Burney, who felt this same motion. Fanny, who watched as seven men in black invaded her study with no warning. They had doctor's bags and her husband's permission. Fanny, who lay there under a handkerchief with nothing but a bit of laudanum in her tea, as they sawed off her breast. When they looked down through her skin and muscle, did they imagine they could see her heart behind its bone bars? They held her down. But, she says, she never fought them.

Again all description would be baffled – yet again all was not over, - Dr. Larry rested but his own hand, & - Oh Heaven! - I then felt the Knife tackling against the breast bone - scraping it! - This performed, while I yet remained in utterly speechless torture, I heard the Voice of Mr. Larry, - (all others guarded a dead silence) in a tone nearly tragic,

The body, for she cannot have a name any longer, is hollowed out, small wet sac by sac. Her gut, her heart, her lungs—all that propelled her—is weighed and bagged. An assistant, whose face we never see, washes her shell carefully. They move her body, and it looks easy; she's light now. Brakhage wants me to see how we can be unburdened; how, then we can be lifted as easily as an empty purse.

Her skin seems so pale and smooth, but I will never know if she was pretty: her scalp remains folded over her eyes and nose like a blindfold.

It takes several tries for me to watch the whole film through.

I wanted to try everything. After Mr. Hamilton's 8th grade biology class dissected cuttlefish, he fried one on a steel plate over a Bunsen burner. I was walking by and a girl dared me to try some. She was being cruel, but I didn't realize that until later. The rubbery texture resisted my teeth; there was no flavor beyond the pickling solution.

Desire everyone present to pronounce if anything more remained to be done; The general voice was Yes, - but the finger of Mr. Dubois - which I literally felt elevated over the wound, though I saw nothing, & though he touched nothing

On some lunch breaks I would visit the chemistry classroom and watch Brutus, the school's moderately socialized python, eat a white rat. Chemistry got him because Mr. Hamilton preferred the docility of fish. On other days, I would ask to hold Brutus, but this opportunity was rare. I loved the paper-dry skin and cool weight of the snake across my shoulders. It pressed, encircled, all one tugging muscle. It was comforting, being held by the snake.

It was probably while wading in the river chasing crawdads that I saw a movement near a tiny pile of twigs and rocks. I lifted the sticks out of the water, but saw nothing, no scale, no antennae or eye. I placed them back in the water and waited until the caddis fly larva stuck two tentative legs out of its assembled shell. I picked it back up, and left it on a smooth river rock to dry. I wanted to watch it crawl back into the water, watch that process of return—as though I were young Darwin with a Galapagos lizard, but instead of hucking it into the sea, I was dragging it out. The larva did not show itself. It did not save itself. Instead, it retreated further into its shell, trying to escape the air that evaporated all movement out of its skin.

There was never any indication, outside of the cluttered confines of my desk top, that a girl could be a scientist. Or that my interests were even scientific. There was no proof that discovery could be a graceful, shining thing.

So indescribably sensitive was the spot – pointed to some further requisition - & again began the scraping!

I learned later, through rumors and shouts and long silent looks not to mention waveforms or snakes or jellyfish. I dropped the microscope in a box and forced myself to stop announcing the names of birds on the long school bus ride home. I noted the way my classmates sat around me and smiled out at the world like open dandelions or full moons. I practiced that.

Years later, when the art college accepted my application, I was relieved to learn that very little science was required. That which intrigued me became too obvious with scalpel and lens at hand. Instead, I drew. My fingernails pressed half-moons into the blue wood of my pencils with a drive to get through the paper, under its skin. I carved deep grooves in linoleum and plaster blocks. I fought every surface, then smoothed wax, glue, and ink over the cuts I'd made. My drawings were of dead grasshoppers, broken toys gone filthy with age, bones, bare roots growing over black, slick rocks: a topographical map to dark cracks in the ground, openings, still and quiet things, that which had been left over and behind.

In the studio I would tell myself, *there's nothing wrong with that.* But I had turned timid and over-careful. A fear had grown inside me where once there had been only fascination. There was no knife, no charged wire loop that could remove it.

I remember reading in a book in college, whose name I have forgotten, about a terrible operation. It was my ex-husband's book: he was very interested in Roman warriors and in Genghis *- and, after this, Dr. Moreau thought he discerned a peccant atom - and still, & still, Mr Dubois demanded atom after atom.* Khan crossing the great swamps of middle Russia. But once I flipped through and found a short letter from a woman who had had breast cancer. She could feel the knife as it bumped across her ribs. She used the word scraping. She could hear and feel it *scraping.* That was over 18 years ago, and I still can't forget her.

It isn't until Internet keywords that I find her again. *scalpel ribs mastectomy historical eyewitness.* Her name was Fanny. She lived for 29 years after the operation. How she must have been grateful for science, and terrified of it, too. How its song and its scream must have rung in her ears.

GENEALOGY // how to skin a bird

Your first and only incision will be right over the sternum. All birds have a bald patch there. Blow lightly on the breast until the feathers begin to part and you can see the pale skin beneath.

Rest your finger there for a moment. Feel the bone your blade will follow. Make a wish, if you must, and then slice from collar to belly carefully.

*

I used to keep the letters my father had written to me in a box with all of my other letters. There were three of them, all written before I was eight, on lined paper with a ripped, spiral fringe. He put them into the envelopes he sent my mother. Otherwise, they were empty, except for a check, always made out for $75. Sometimes the envelopes came from Alaska or Tahiti.

When I was a little older, if I saw the mail before my mother, I would feel the envelopes, to see if they were thicker than just-a-check.

You will need to cut the spinal column and trachea before you go any further. To do this, carefully work a small surgical probe between the skin and neck muscles at the very top of your cut. The curved tip is blunt, and if you advance it slowly, it shouldn't tear the skin. Work it behind the neck until you can see it on the other side. Then slide the lower blade of a scissor along the steel. When you can see the scissor's point, cut.

On larger birds, you may need to cut two or three times, blindly. This is why the probe is important: it keeps you from cutting through the back of the neck and beheading your specimen.

*

I stayed at my father's parent's house a few times as a child, usually when he was visiting, too. Even though he had a local apartment twice, I never stayed at either one. In the second apartment—which he took me to once for just a moment—I had my own room. I looked in through the darkened doorway and saw a bed and bedside table, no lamp, and then we had to leave.

I never stayed alone with him until I was in college. This doesn't feel ominous, just disconnected.

His parents lived in a double wide trailer, in the yard of a landlord I never met. I slept in his little brother's old room.

After the neck is cut, you will slip the wings from their bones like a jacket.

To do this, repeat the following steps for each side: using your fingers, pinch the humerus below the shoulder joint and slide the skin away from the muscles like you are taking off tight jeans or panty hose.

Keep rolling the skin from the structure until you feel a boniness branch off from the ulna, just past what would be

14 | Chelsea Biondolillo

the elbow joint. This is the first primary feather. Use your fingernail to scrape it out—as you would a wasp's stinger in your own arm. There will be a few primaries. Pop them from the bone one by one, and then cut the humerus close to the shoulder. Be careful not to cut through the skin. Both wings should be intact in the final specimen.

*

Uncle John had hundreds of comic books. I lay in that strange bed and read them through hours of sleeplessness. I preferred Casper and Archie to the superheroes because of the simplicity of their drawings. *Hulk* and *Iron Man* comics seemed busy and cramped; I couldn't always tell what was happening.

My grandmother, his mother, was an Avon lady; she sometimes let me take home tiny lipstick samples of last season's colors. She had an impressive collection of facsimile whiskey decanters shaped like famous people and a kangaroo skin my father had brought her from Australia.

My grandfather had a wooden leg and a glass eye. I was not told this, but discovered the prosthetics while snooping. I was told that used to drink homemade corn liquor that he filtered through loaves of coarse wheat bread. He did this a long time before I was born. I don't remember who told me that.

When both wings are free from the body, gently peel the back skin from the muscle. If it sticks, as can be the case with some freezer-burned birds, use the probe and scalpel to loosen or cut through the scab.

You will cut the legs by rolling the skin away from them as you did with the wings, but stop sooner, before you get to the scaled leg skin. Cut the thigh bone even closer to the body than you did on the wings.

Now pinch the tail ligaments and cut just above where you can feel the feathers end. If you cut too high, it will be a bit messy, but not unsalvageable. If you cut too low, the tail will fall off. This can't be fixed, so err on the side of a mess.

*

My father's mother would fry trout in thick oil for dinner.

We ate peanut butter and mayonnaise on white bread for lunch. I asked that this sandwich be put in my bagged lunch, when I got back home to my mother.

For breakfast, my grandmother would make tiny pancakes for my father and me. He would put butter and jam on them, instead of syrup. He would roll them up like cigars and eat them with his hands. So would I.

The body, separated, should resemble a very small version of a fryer or turkey. Notice that the muscle tissue of songbirds, however, is all dark meat.

You will carefully collect tissue samples from the breast, heart, and liver now, and determine the sex of the bird. You may also need to collect stomach contents.

Follow the photocopied instructions for this step. Use the small plastic vials, and consider all samples dry unless your collections manager tells you to store them wet, as is sometimes the case with rare specimens.

*

My mother never spoke ill of him, I want to say that. But the spaces between his contact swelled with meaning by the time I got to high school. First his father died, and then his mother. I asked him if I could come to her funeral, after I learned of it from my mother's mother—her neighbor. I smoked Marlboro Lights in front of him for the first time that weekend.

I stayed with him and one of the women he married a couple of times, once in Spokane and once in Pleasant Hill, California. She had older children, cut his meat for him, and talked in a nasal, baby voice.

He didn't travel overseas anymore, but would frequently overspend. During one visit, he bought a new car and a giant television. Then, he got a nosebleed and sat in the center of his bed staring into space. His wife said, "Go comfort him!" Their marriage may have lasted ten years. Maybe less.

The head is difficult, so take your time.

Start by pinching the neck and working the skin away from and over the back of the skull. When you get to the eye sockets, use your scalpel to carefully cut through the thin membrane that connects the lids to the eyes.

You will need to keep the majority of the skull intact, to maintain a natural shape, but the soft tissue should all be removed: tongue, eyes, brain. Use caution so that when you cut out the tongue, you do not damage the hinges that connect the upper and lower maxilla.

*

The last time I stayed at my father's house, he was living alone in a near-bare apartment outside of Alameda, near a marina. We ate out, because the only thing in his refrigerator was a whole roasted chicken from the grocery store. He lamented my inability to cook. He showed me his cupped palm, full of old wedding bands. My mother's was not there, he said, as he clamped his hand closed and carried them back to wherever they'd been hidden. Did he think I might have asked for it? It broke, he said, making a joke of it.

I awoke the next morning in his (only) bed with his arm around me. I spent the rest of the week on the living room floor without comment.

The specimen will be used for study, it isn't intended to seem life-like—keep that in mind as you finish. After washing and drying the skin, you will use cotton wrapped around a small dowel to replace what you have extracted. Use enough cotton to create a sense of fullness, in the eye sockets, in the belly.

Tie the legs together, so they do not splay, and attach the specimen tag using a square knot. Paint the ends of the string with clear nail polish. Then close up the original incision.

*

After college, after my own failed marriage, he regained an interest in me. I tried to be polite when he came to visit me in Washington DC. I met him downtown, but did not invite him to my house.

A few years later, one of his girlfriends called me to say he was depressed and would I please call back later to cheer him up? He'd mentioned her once. She had a little boy, he'd said, whose name was spelled 'Jesus.'

"It's pronounced 'hey-Zeus'," he'd said. I know, I thought.

He had told me her name, but I'd forgotten it and so had hung up on her the first time she called, thinking it was a wrong number. She spoke very little English. The conversation was frustrating for us both. I didn't call back.

You may be tempted to take your needle and thread and sew dozens of the smallest stitches you can, as though attention to detail could hide the hole you've made. Resist this urge. Everyone expects a hole in an empty skin. Dip your needle once twice three times and then pull the thread tight. It's enough.

*

When I stopped answering his emails, he would telephone. The calls came every year or so, between girlfriends or wives. Once, late at night in a Barcelona train station, I was confused by the time difference and answered my cell phone, thinking there must be an emergency. He was bored and lonely. He made a joke about how long my master's degree was taking.

When I stopped answering the phone, he started to text. He always tells me first, that he's broke, and second, that he's single.

Before you position the bird to be dried, stick one straight pin through the base of the bill vertically from bridge to chin. Thread a small piece of twine through the nostrils and around the bill, using the pin as an anchor to cinch it tightly closed.

Use as many pins as you need to hold the specimen in exactly the position you want on the board. The head should be facing forward or slightly up, the wings tightly folded, right leg over left, and tail feathers spread just to the width of the body. With proper preparation, it will keep this shape through decades of routine handling.

PYROLOGY // an account(ing)[*]

On the sidewalk, at dusk: the head of a razor, a plastic filter from a cigarillo, and other unconnected, broken pieces of lives.[1]

122°F on June 26, 1990;

121°F on July 28, 1995;

120°F on June 25, 1990;

119°F on June 29, 2013;

118°F on July 16, 1925;

118°F June 24, 1929;

118°F July 11, 1958;

118°F July 4, 1989;

118°F June 27, 1990;

118°F June 28, 1990;

118°F July 27, 1995;

118°F July 21, 2006;

118°F July 2, 2011;

I suppose no man ever saw Niagara for the first time without feeling disappointed. I suppose no man ever saw it the fifth time without wondering how he could ever have been so blind and stupid as to find any excuse for disappointment in the first place. I suppose that any one of nature's most celebrated wonders will always look rather insignificant to a visitor at first, but on a better acquaintance will swell and stretch out and spread abroad, until it finally grows clear beyond his grasp - becomes too stupendous for his comprehension.

---

[*] *Includes text from Mark Twain published in The Sacramento Daily Union, November 16, 1866.*

[1] *I sat outside in the steam through monsoon rains, counting thunder and lightning, back when I was a smoker, back when I was a wife.*

Ticking, like the stone clocks under every mountain. Breaking, like moonlight through someone else's beige blinds. A hum, hard to place.[2]

1991 – Mount Pinatubo;

1980 – Mount St. Helens;

1912 – Novarupta;

1902 – Santa María;

1886 – Mount Tarawera;

1883 – Krakatoa;

1835 – Cosigüina;

1815 – Mount Tambora;

1783 – Grímsvötn;

1650 – Kolumbo, Santorini;

1600 – Kuaynaputina;

1580 – Billy Mitchell;

1477 – Bárðarbunga;

1452 – Kuwae;

1280 – Quilotoa.

The greater part of the vast floor of the desert under us was as black as ink, and apparently smooth and level; but over a mile square of it was ringed and streaked and striped with a thousand branching streams of liquid and gorgeously brilliant fire! It looked like a colossal railroad map of the State of Massachusetts done in chain lightning on a midnight sky. Imagine it - imagine a coal-black sky shivered into a tangled network of angry fire!

---

2   *It doesn't do any good to remember the exact smoothness of your shoulder; but it snowed last night and my room is cold.*

Every spring fails by autumn. What I'm saying is that the problem is with the word itself.[3]

2005 – M_____;

2012 – T_____;

2008 – Y_____;

2000 – A_____;

1990 – S_____;

1992 – A_____;

1994 – D_____;

1993 – A_____;

1991 – B_____;

1992 – M_____;

2011 – J_____;

2010 – J_____;

2009 – R_____;

1989 – J_____;

1987 – P_____.

I forgot to say that the noise made by the bubbling lava is not great, heard as we heard it from our lofty perch. It makes three distinct sounds - a rushing, a hissing, and a coughing or puffing sound; and if you stand on the brink and close your eyes it is no trick at all to imagine that you are sweeping down a river on a large low pressure steamer, and that you hear the hissing of the steam about her boilers, the puffing from her escape pipes and the churning rush of the water abaft her wheels. The smell of sulfur is strong, but not unpleasant to a sinner.

---

3   *It is easy to box your things: letters, photos, mix tapes. But now I see you in the lines around my eyes—where can I put those?*

NECROLOGY // raccoon, pronghorn, mule deer, ring-necked pheasant, fox

I'm sorry for driving past and driving past and driving past all winter and into spring, and for watching, with interest— even, I'm ashamed to say, a kind of gross curiosity—as you became less and less of what you were, as you were ground down by innumerable tires into bone, fur, and dirt, as you were picked apart by magpies and crows.

I would like to be the kind of person who looks away from the slumped backbone, the twisted leg, the handful of feathers, flickering without flight in the gusts of dusty farm-to-market traffic. But I'm afraid I'll always stare.

Acknowledgements

Phrenology first appeared in *Hayden's Ferry Review*, 53
Fall 2013

(Genealogy) How to Skin a Bird first appeared in *Shenandoah*,
63.2 Spring 2014

Pyrology first appeared in *Sonora Review*, Winter 2014

Raccoon, Pronghorn, Mule Deer... first appeared in *PEN
America/Guernica*, July 2014

Biography

Chelsea Biondolillo is the author of #*Lovesong* (Etchings Press) and *The Skinned Bird* (Kernpunkt Press, 2019). Her prose is collected in *Best American Science* and *Nature Writing 2016, Waveform: Twenty-first Century Essays by Women, How We Speak to One Another: An Essay Daily Reader* and others. She has an MFA from the University of Wyoming and currently lives about 30 miles outside of Portland, Oregon, where she is a regulatory analyst by day and a writer by night.

Colophon

Titles font in Bookman Old Style.

Body text font is Sabon.

Inset boxes font is IM FELL English Pro.

Etchings Press

Etchings Press is a student-run publisher at the University of Indianapolis. Each year, student editors choose the Whirling Prize, a post-publication award, in the fall and coordinate a publication contest for one poetry chapbook, one prose chapbook, and one novella in the spring. For more information, please visit etchings.uindy.edu.

Previous winners and publications

Poetry
2019: *As Lovers Always Do* by Marne Wilson
2018: *In the Herald of Improbable Misfortunes* by Robert Campbell
2017: *Uncle Harold's Maxwell House Haggadah* by Danny Caine
2016: *Some Animals* by Kelli Allen
2015: *Velocity of Slugs* by Joey Connelly
2014: *Action at a Distance* by Christopher Petruccelli

Prose
2019: *Dissenting Opinion from the Committee for the Beatitudes*
        by Marc J. Sheehan (fiction)
2018: *The Forsaken* by Chad V. Broughman (fiction)
2017: *Unravelings* by Sarah Cheshire (memoir)
2016: *Pathetic* by Shannon McLeod (essays)
2015: *Ologies* by Chelsea Biondolillo (essays)
2014: *Static: Stories* by Frederick Pelzer (fiction)

Novella
2019: *Savonne, Not Vonny* by Robin Lee Lovelace
2018: *Edge of the Known Bus Line* by James R. Gapinski
2017: *The Denialist's Almanac of American Plague and Pestilence*
        by Christopher Mohar
2016: *Followers* by Adam Fleming Petty